VOLCANO ADVENTURE

WRITTEN AND ILLUSTRATED BY
STU DUVAL
COVER DESIGN BY
PAUL XU AND STU DUVAL
Art Director:
PAUL XU
Editor:
Jon Thibault

MOTHER NATURE LTD.

Copyright © 2008 by Mother Nature Ltd.

1411 4th Avenue, Suite 765
Seattle, WA98101

All rights reserved. No part of this
publication may be reproduced, stored in a
retrieval system, or transmitted by any
means, electronic, mechanical,
photocopying, recording or
otherwise, without the prior permission
of the publisher

ISBN-13：978-0-9814547-5-7
ISBN-10：0-9814547-5-5

Printed in Taiwan

Characters

Ranger Redd

Park Ranger and adventurer. From his base at Ranger H.Q., Ranger Redd protects and defends the enviroment.

Kate

Loves the outdoors and exploring nature. Kate's hobbies are reading, hiking and computers.

Kimo

Loves chess and video games. Not keen on hiking. Best friends: Kate and Dick.

Dick

Loves photography, but he takes more chances than pictures! Hobbies are rock collecting and climbing.

VOLCANO ADVENTURE

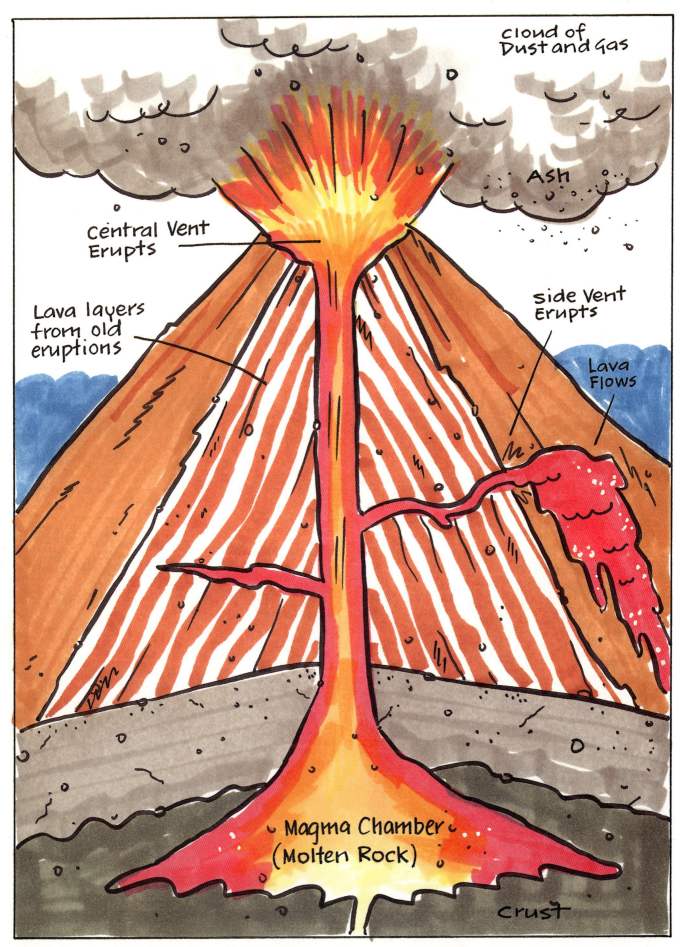

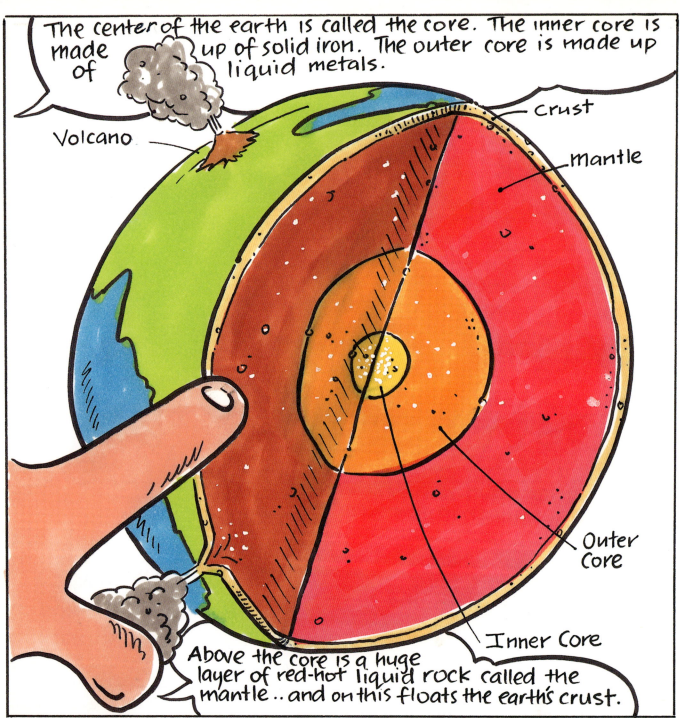

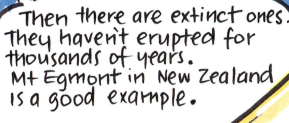

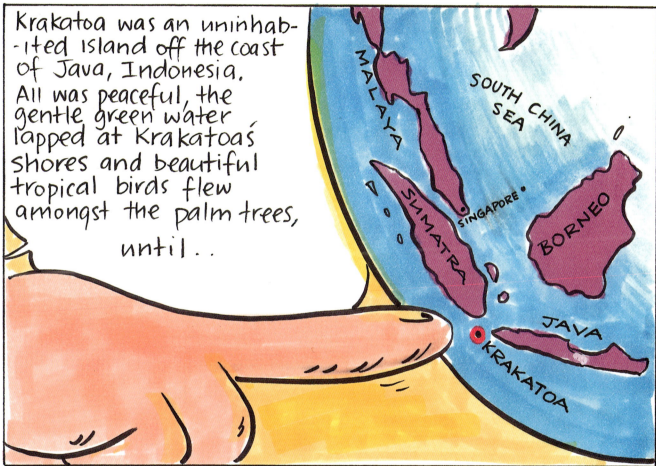

When the volcano wall collapsed, sea water poured into the hot magma below! This caused super-heated steam to billow out...

"Then what!?"

Then came the final explosion — an eruption so loud, it was heard 3000 miles away!! Rocks were hurled 200 ft into the air!

And then...

An enormous tidal wave caused by the eruption destroyed hundreds of nearby villages. 36,000 people were drowned!

For 4 whole days, Mt. St. Helens belched 30 million tons of ash into the atmosphere! The eruption caused the largest landslide ever recorded. It flattened forests and living creatures for 300 sq. miles.

PUMICE

Pumice is formed when lava cools and traps air bubbles in a light porous stone. Pumice is light enough to float on water.

BASALT

Basalt is a type of rock that is formed from dark, runny lava with tiny crystals.

GRANITE

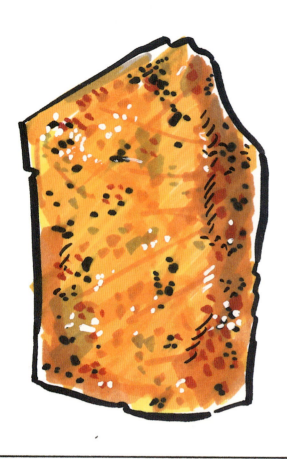

Granite is a rough, grainy, volcanic type of rock formed when magma cools in the crust. It contains large crystals of quartz and mica.

Cool! I can see tiny mineral crystals in this basaltic rock! But not all these rocks came from liquid lava.

Some volcanos just shoot out solid pieces of rock... they're often called Volcanic Bombs!!!

The tiny bits are volcanic ash.

Finally, Dick sleeps...

But Vulcan does not!

Deep within the volcano, magma, ash, rock, and gas are under pressure! And it's looking for a way out!

Central Vent

Magma

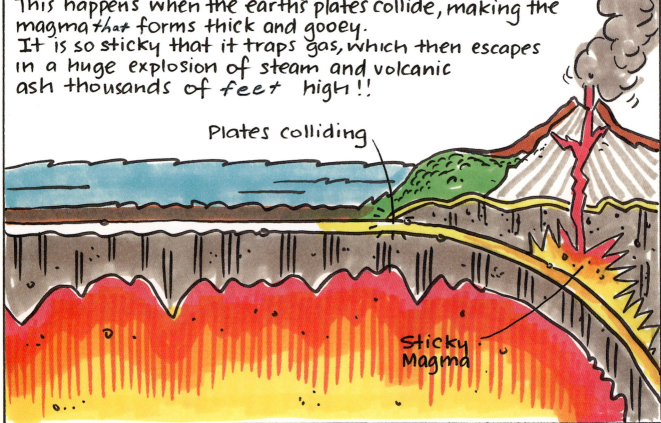
This happens when the earth's plates collide, making the magma that forms thick and gooey.
It is so sticky that it traps gas, which then escapes in a huge explosion of steam and volcanic ash thousands of *feet* high!!

Plates colliding

Sticky Magma

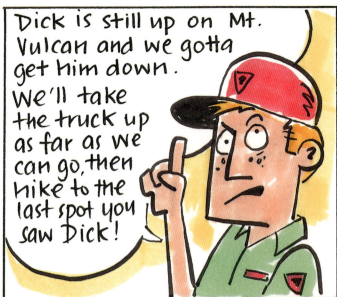

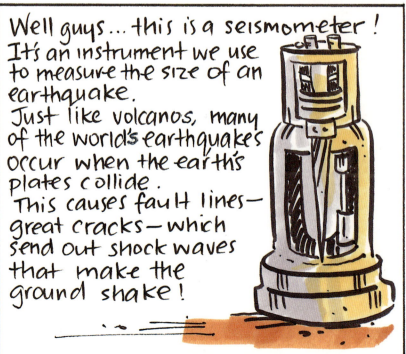

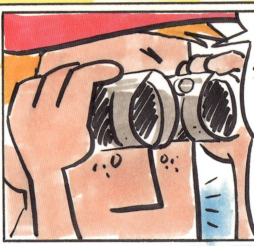

With one ear-splitting eruption, Mt. Vulcan explodes! It releases all the gases dissolved in the magma. Lava and ash are belched out in clouds of steam. The molten lava can reach temperatures of over 1,000°C! It will burn and flatten anything in its path!

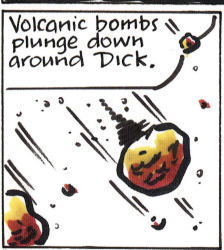

There are over 2,500 geysers in Yellowstone Park in the U.S.A. Rotorua, in New Zealand, recorded the tallest geyser ever. It was called Waimangu, and in 1903 spouted.

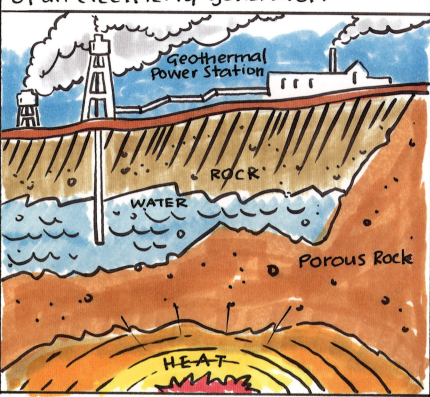

The steam can be piped directly from the ground and piped to power stations, where it turns the turbines of an electricity generator.

So, in a way, it's the volcano that provides the power we use!

If warm water mixes with soil and chemicals underground, it produces hot mud pools.

Amazing Volcano Facts

Did you know that there is a volcano off the coast of Italy that is called the lighthouse of the Mediterranean because it erupts every 20 minutes?

Did you know that just before a huge earthquake in Haicheng, China, in 1975, hundreds of snakes were seen leaving their burrows They could sense the vibrations before the shake!

Volcanologists

Scientists who study and measure volcanoes
are called volcanologists.
Their job is both exciting and highly dangerous!
Many volcanologists have been killed suddenly when a
volcano they were studying erupted.
They wear special heatproof suits to protect them when
observing volcanoes up close.

Volcano Adventure Quiz

Test yourself with these questions. All the answers are to be found in this book.

1   What is the name of the most violent eruption in modern times that was so loud it was heard 500 km away?

2   What was the name of the brand new volcanic island that suddenly appeared off the coast of Iceland in the 1960's?

3   What are the large pieces of rock that shoot out of an eruption called?

4   What is the name of the instrument used to measure the size of an earthquake?

5   What planet has the largest volcano in the solar system? Do you know the volcano's name?

ANSWERS 1 KRAKATOA 2 SURTSEY ISLAND 3 VOLCANIC BOMBS 4 SEISMOMETERS 5 MARS, MT OLYMPUS

Glossary

ACID RAIN Acid rain is caused by sulphur dioxide and nitrogen oxides in the air.

ACTIVE A volcano that is likely to erupt and is not extinct.

ATMOSPHERE Gases which surround the Earth.

BASALT Dark rock formed from solidified lava flows.

CORE The center of the Earth made up of solid iron. Around it is the outer core made up of hot liquid metals.

CRATER Bowl-shaped mouth of a volcano.

CRUST The outer layer of rock around the Earth.

DORMANT A volcano that is live but has not erupted.

EARTHQUAKE The shaking of the Earth's surface caused by faults or volcanic activity.

ERUPT Rocks, gases and lava forced up through a volcano's vents.

GEOTHERMAL A heat source found deep inside the Earth.

GEYSER An underground spring of hot water that spurts up in a stream of steam to the surface.

GRANITE A crystal-studded volcanic rock.

LANDSLIDE A mass of earth and rock sliding down a mountainside.

LAVA Hot, liquid rock (magma) which has reached the Earth's surface.

MAGMA The hot liquid rock inside the Earth.

PLATE One of the massive sections of rock on the outer layer of the Earth.

POROUS ROCK Rock through which water can flow.

PUMICE Light porous kind of rock.

SEISOMETER An instrument that measures movements in the Earth.

VENT An opening in the side of a volcano through which lava flows.

VOLCANIC BOMB Large pieces of rock that shoot out from an erupting volcano.

VOLCANOLOGIST A scientist who studies volcanoes.